Susan L. Roth and Cindy Trumbore

Butterfly
for a
King

Saving Hawai'i's Kamehameha Butterflies

collages by Susan L. Roth

Lee & Low Books Inc. *New York*

Remembering Claude Freychet —*S.L.R.*

For Darcy, a beautiful butterfly —*C.T.*

A spiny caterpillar becomes a magnificent butterfly. The butterfly lays eggs that become spiny caterpillars. The cycle goes on and on. But what happens when the cycle is broken? What if the butterflies start to disappear?

This is the story of a beautiful butterfly that lives in Hawai'i and nowhere else on Earth. It is also a story of citizen science in action, where ordinary people worked with scientists to help save the butterfly.

And it all started with a bang.

Lava flowed . . .

Millions of years ago, melted rock blasted up from a searing hot spot in the floor of the Pacific Ocean. The melted rock, called lava, hit the water and cooled, forming a volcano. Over time, layers of lava piled up. The lava grew above the surface of the water and spread out to form an island.

and islands rose.

The volcano sat on a plate, a moving part of Earth's outer crust. As the plate moved very, very slowly over the hot spot, at about the speed fingernails grow, new volcanoes and new islands formed.

Islands rose . . .

As the islands rose one by one, they formed the chain
that is known today as Hawai'i.

and formed a chain.

The nearest continent was more than 2,000 miles
(3,219 kilometers) away.

Plants reached the chain . . .

Some plants blew to the islands as seeds. Other plants floated there on the water. Still others probably arrived as seeds in the bellies of birds. Tens of thousands of years could pass before each new plant arrived.

and a butterfly landed.

Many plants were good hosts for the bees, butterflies, moths, and other small animals that landed in Hawaiʻi. And a few of the plants were good hosts for the ancestors of one special butterfly that arrived in Hawaiʻi from North America or Asia long ago.

The butterfly landed . . .

This special butterfly is found only in Hawai'i, where it flies over mountains and along streams and valleys. The adult butterfly drinks nectar from flowers and sap from trees, especially the koa tree.

koa (KOH-ah)

and years sped by.

When the butterfly folds its lower wings over its upper ones, it looks like tree bark. When it unfolds its wings, it shows its fuzzy brown body and beautiful orange, black, and white markings.

Years sped by . . .

Halley's Comet was first observed in the sky more than
two thousand years ago. It zips around the sun and is visible
from Earth about every seventy-five years.

In 1758, Halley's Comet was seen in the night sky above

and a comet blazed.

Hawai'i. But even before this sighting, stories about a comet were told on the islands. People said that a light with feathers like a bird would appear in the sky and then a great leader would be born.

The comet blazed . . .

No one knows for sure when the great leader—King Kamehameha—was born, but some historians suggest 1758 because Halley's Comet was seen that year.

Until the early 1800s, the islands were ruled by warring chiefs. Kamehameha was a tall, strong warrior from the island of Hawaiʻi, which would give its name to the whole

Kamehameha (ka-MEH-ha-MEH-ha)

and a king was born.

chain. He battled the rival chiefs to become king of all
Hawai'i in 1810.

   In time, the special orange, black, and white butterfly was
named the Kamehameha butterfly to honor the king who
brought all the islands together under one rule.

Children loved the king . . .

Hawai'i

Today Hawai'i is one of the fifty states of the United States.
The people who live in Hawai'i still remember the great King
Kamehameha, and they love the butterfly named for him.

In 2009, six fifth-grade students in Hawai'i thought there
should be an official state insect. Naturally, they considered the
Kamehameha butterfly. They also thought about Hawai'i's happy
face spider (even though spiders are not insects). Luckily for

and children spoke up.

the Kamehameha butterfly, the students decided people like butterflies more than spiders.

With their teacher's help, the students prepared talks, wrote letters to lawmakers, and then went to the state capitol in Honolulu. They asked the state's leaders to pass a law saying that the Kamehameha butterfly was the state insect.

Children spoke up . . .

The students told lawmakers that the butterfly's numbers were shrinking. They hoped that recognizing the butterfly as the state insect would make people want to protect it and help it survive. The students pointed out that the butterfly was named for King Kamehameha, who had united their islands. And the butterfly helped pollinate plants. This

and a law was passed.

was like sharing *aloha*. Aloha is a traditional Hawaiian word used to say hello and good-bye, but it means much more. When you practice aloha, you remember always to have respect and compassion for others and to give joyfully.

The state lawmakers agreed with the students. They voted to pass a law that made the Kamehameha butterfly the state insect.

aloha (ah-LOH-hah)

The law was passed . . .

As residents of Hawai'i learned that the Kamehameha butterfly had been named the state insect, they realized they weren't seeing the butterflies around much anymore. Animals and insects that had been brought to the islands from other

and people asked questions.

places were destroying them. There weren't as many of the plants the butterflies needed to live as there used to be. People contacted the state government and asked if anyone was helping the Kamehameha butterfly.

People asked questions . . .

To help save the Kamehameha butterfly, the state's Department of Land and Natural Resources worked with the University of Hawai'i. Together they created the Pulelehua Project. *Pulelehua* is the Hawaiian word for *butterfly*. The word also means "to be scattered, as if being blown by the wind"—a good description of how butterflies fly.

pulelehua (POO-leh-leh-HOO-ah)

and a project began.

First the project's scientists had to understand where the butterflies were still found on the islands. They asked the people of Hawai'i for help collecting data. Through a website, the scientists encouraged volunteers, called citizen scientists, to send in photos of Kamehameha butterflies in all stages of their life cycle. With the photos, people included information about when and where each stage was seen.

The project began . . .

Female Kamehameha butterflies lay their eggs on the top or bottom of the leaves of certain trees and bushes. They choose only plants with leaves that their caterpillars can eat.

and butterflies laid eggs.

The most common of these plants is māmaki, a plant native to the Hawaiian Islands. So citizen scientists searched plants for eggs that look like tiny brown or gold jewels.

māmaki (MAH-mah-kee)

Butterflies laid eggs . . .

Citizen scientists also looked for Kamehameha caterpillars that had hatched from their eggs.

Using sharp mouthparts called mandibles, a caterpillar cuts a half circle in the edge of a leaf. Then the caterpillar pulls the

and caterpillars hatched.

cut part of the leaf over itself and seals it up with silk it spins with its mouthparts. This forms a shelter like a little tent. Citizen scientists looked for these shelters too.

Caterpillars hatched . . .

A growing caterpillar munches on its shelter, sometimes creeping out to feed elsewhere on the leaf. When the shelter has too many holes, the caterpillar moves and makes a new shelter. As it eats and grows, the caterpillar gets too big for its skin. So it sheds its skin, showing new skin underneath.

and shed their skins.

After shedding its skin four times, the caterpillar is bright green with little spines. Then it hangs upside down from a leaf or twig and sheds a fifth time, transforming to a chrysalis. Citizen scientists searched for these chrysalises dangling from leaves and twigs.

chrysalis (KRI-suh-liss)

Five skins were shed . . .

Inside the chrysalis, a Kamehameha caterpillar's body slowly turns into a butterfly. After ten to fifteen days, the adult butterfly emerges from its dangling chrysalis, flexes its wings, and flies away looking for food. It uncurls its tonguelike proboscis to sip nectar from flowers and sap from koa trees.

proboscis (proh-BOSS-kiss)

and butterflies were born.

Female Kamehameha butterflies have white spots at the tips of their wings. Males have both light-orange and white spots on their wingtips. So citizen scientists looked for the bright patterns of the butterflies' wings.

Butterflies were born . . .

While citizen scientists were collecting data, the project's scientists were breeding Kamehameha butterflies in an insect lab. They tried different ideas to get the butterflies to lay eggs. One scientist read that a relative of the Kamehameha butterfly liked to mate at sunset. So he brought captive butterflies to the lab's roof to see the beautiful Hawaiian sunsets. Success! The butterflies mated and the females laid eggs.

and scientists tried new ideas.

The project's scientists had a computer program analyze all the data that had been collected. Then they used the results to make a map showing the best places to release the butterflies they had raised in the lab.

To get ready for the releases, people planted māmaki and other plants that would attract Kamehameha butterflies.

Scientists tried new ideas . . .

In 2017 the hopes of the students who wanted to help the Kamehameha butterfly eight years earlier were fulfilled.

On a beautiful day in April—just before Earth Day—people from the Department of Land and Natural Resources took a carrier filled with ninety-four butterflies to a marsh on the island of O'ahu. They put sugar water on their fingers, and butterflies

O'ahu (oh-AH-hoo)

and butterflies flew.

climbed on. Then they held their hands in the air and watched the butterflies fly away.

Releases of thousands of butterflies followed in other areas. One of the most promising release sites, the Mānoa Cliff Restoration Area on Oʻahu, is cared for entirely by devoted volunteers who have been restoring native plants there for many years.

Mānoa (MAH-no-uh)

Scientists in Hawaiʻi continue to look for new ways to help Kamehameha butterflies survive. The scientists have begun releasing eggs in the wild instead of butterflies. Eggs are easier to raise in the insect lab, and the scientists can release thousands in just one week. They put the eggs in empty paper tea bags and clip the tea bags onto māmaki leaves. When the eggs hatch, the caterpillars climb out of the tea bags and right onto the leaves.

A spiny caterpillar becomes a magnificent Kamehameha butterfly. The butterfly lays eggs that become spiny caterpillars. And with the help of many hands, the cycle goes on and on.

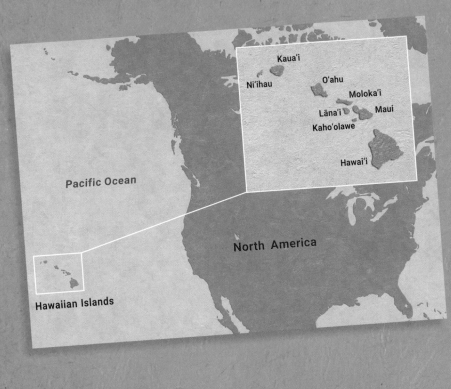

Statue of King Kamehameha in front of Hawai'i's State Supreme Court building, Honolulu

Afterword

Hawai'i has eight main islands and many smaller ones that together stretch about 1,500 miles (2,414 kilometers) across the Pacific Ocean. Scientists once thought the islands were settled over hundreds of years, starting sometime between the years 300 and 600. More recent research suggests that the islands were settled later and in a shorter time period: between about 1219 and 1266.

The first people to reach the islands traveled in canoes from islands in West Polynesia thousands of miles away. Several different kingdoms existed across the chain until King Kamehameha (ca. 1758–1819) unified the islands into a single kingdom in 1810. His reign is celebrated every June 11 in Hawai'i as a state holiday featuring parades and ceremonies. Statues of the king are decorated with colorful wreaths of flowers called lei.

In the Kumulipo, the sacred Hawaiian creation chant that tells how life began on the islands, the butterfly is one of the first creatures named. The Kumulipo includes these lines about the Pe'elua (caterpillar):

The Pe'elua was born and became parent;
Its offspring was a flying Pulelehua.

Ancestors of the Kamehameha butterfly arrived in Hawai'i millions of years ago. The butterfly was first described by scientist Johann Friedrich von Eschscholtz in 1821. Its scientific name is *Vanessa tameamea*.

The fifth-grade students who asked to have the butterfly named the state insect of Hawai'i in 2009 attended Pearl Ridge Elementary School on the island of O'ahu. They said the butterfly represented the beauty and history of their islands and the unity among them.

lei (ley)
Kumulipo (koo-moo-LEE-poh)
Pe'elua (peh-eh-LOO-ah)
Vanessa tameamea (vah-NESS-ah tah-MEH-ah-MEH-ah)

Butterfly egg; a female butterfly can lay up to 300 eggs in its lifetime

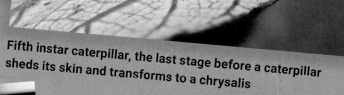

Fifth instar caterpillar, the last stage before a caterpillar sheds its skin and transforms to a chrysalis

Chrysalis from the Pulelehua Project hanging from the lid of a cup

Adult Kamehameha butterflies: a female has white spots on its wings; a male has both white and orange spots

Pearl Ridge Elementary School students with state representatives

There are millions of species of insects in the world, and they are disappearing from Earth at an alarming rate. Climate change, habitat loss, the introduction of non-native predators, and the use of pesticides all contribute to declines of insect species; and when insects disappear, the animals that eat them and the plants they pollinate also suffer. Some of these conditions contributed to the decline of the Kamehameha butterfly, the first Hawaiian insect to be the subject of a captive-breeding and reintroduction program.

Hawai'i's Department of Land and Natural Resources (DLNR), which protects and preserves the state's incredible range of insects, plants, and animals, provided the funding for the Pulelehua Project. The DLNR's Division of Forestry and Wildlife includes a program that studies Hawai'i's invertebrates, animals that do not have backbones, including all insects.

From egg to adult, a butterfly only lives for about 45 days

Cynthia B. King, who coordinates the state's Hawai'i Invertebrate Program, approached entomologist Dr. Will Haines to lead the reintroduction efforts for the Kamehameha butterfly. The project now has a dedicated insectary for breeding the butterflies and other rare insects.

Dr. Will Haines at the insectary

Growing butterflies at the insectary

Butterfly born at the insectary

The project also uses a software program to analyze data and create a distribution map. The map shows where Kamehameha butterflies have been seen. It also indicates the places in Hawai'i that have living conditions—plants, temperatures, sunshine, rainfall, and other resources—the butterflies need to survive. With this information, scientists can pinpoint areas where reintroductions are most likely to be successful. Scientists count a release a success if afterward they spot butterflies and caterpillars in a native habitat where the insects had once disappeared.

Koa tree; its sap provides food for Kamehameha butterflies

Māmaki; female Kamehameha butterflies lay eggs on the plant's leaves

Butterflies at the Honolulu Zoo; a sweet sports drink on the towel provides nutrition

As the project continues, the scientists try new techniques to help the butterflies survive and reproduce. The Honolulu Zoo has its own population of Kamehameha butterflies, and the scientists learn from the work done at the zoo too.

The Pulelehua Project still relies on citizen scientists for sightings of the Kamehameha butterfly in all stages of its life cycle. If you live in or visit Hawai'i and spot a Kamehameha butterfly, caterpillar, egg, or chrysalis, you can become part of the project by taking a photo. Add a description of the sighting location and the date, and then visit the Pulelehua Project webpage (https://www.inaturalist.org/projects/pulelehua-project) for information about how to submit your photo.

Mahalo—thank you!

Illustrator's Note

I have been completely humbled by the Kamehameha butterfly: by its beauty and its precision and its intricacies.

This is how I made the butterflies for the pages of this book. I started by assembling the following: two very good color photographs of the Kamehameha butterfly (one female, one male); two pairs of sharp scissors (one very small with thin, slightly curved blades and one a little bit larger with pointy, straight blades); one pair of sharp, pointy tweezers with tips that meet precisely; colored papers (black, orange, white, and brownish-orange); and double-sided adhesive tape. Then I studied the photographs for all the details they provided. After that I very carefully cut out the shapes I needed in each color and arranged them to form butterflies. Last, I used the adhesive tape to stick the cut pieces of colored papers together. KAMEHAMEHA BUTTERFLIES!

It is with amazement and in total awe of the real butterflies that I ask you to accept my images as pretty close representations. But if you want to see these beautiful butterflies as they really are, you'll just have to visit them in Hawai'i! —S.L.R.

Acknowledgments

The authors would like to thank the following people for their generous help with this book: The Pulelehua Project's Dr. Will Haines and Hawai'i Division of Forestry and Wildlife Entomologist Cynthia B. King, who read the text for scientific accuracy, granted interviews, and provided photographs; Laura Debnar of the Honolulu Zoo, who shared information about the zoo's Kamehameha butterfly display; Adriane Awaya, parent community network coordinator for Pearl Ridge Elementary School, for the photo of the children who started it all; and Sam 'Ohu Gon III, our cultural consultant, who works at The Nature Conservancy of Hawai'i.

S.L.R. also thanks Anne Athanassakis, Sharon Cresswell, Olga Guartan, Sabra Black Hoffman, Christine Kettner, Nancy Patz, Kale Taylor, and JR et al.

Authors' Sources

Del Pino, Brittany Moya. "Cracking the Butterfly Code." Wired, March 19, 2015. https://www.wired.com/2015/03/cracking-the-butterfly-code/.

Gorelick, Glenn A., and Ronald S. Wielgus. "Notes and Observations on the Biology and Host Preferences of *Vanessa tameamea* (Nymphalidae)." *Journal of the Lepidopterists' Society* 22, no. 2 (1968): 111–114. http://images.peabody.yale.edu/lepsoc/jls/1960s/1968/1968-22(2)111-Gorelick.pdf.

Haines, Dr. Will. Personal interview with Cindy Trumbore, January 14, 2019.

Jarvis, Brooke. "The Insect Apocalypse is Here." *The New York Times Magazine*, December 2, 2018, New York edition.

"Kamehameha Butterflies Return to O'ahu Forests This Earth Day." Hawai'i Department of Land and Natural Resources, April 21, 2017. https://governor.hawaii.gov/newsroom/latest-news/dlnr-news-release-kamehameha-butterflies-return-to-oahu-forests-this-earth-day/. Includes video "The Butterfly Effect." https://vimeo.com/214227025.

"Kamehameha Butterfly Considered as State Insect." Hawai'i House of Representatives Blog, February 28, 2009. http://hawaiihouseblog.blogspot.com/2009/02/kamehameha-butterfly-considered-as.html.

King, Cynthia B. Personal interview with Cindy Trumbore, January 15, 2019.

Ladao, Mark. "Reviving the Kamehameha Butterfly Population at Tantalus." Ka Leo, February 5, 2018. http://www.manoanow.org/kaleo/news/reviving-the-kamehameha-butterfly-population-at-tantalus/article_39134810-0a26-11e8-8b05-d318d2728572.html.

Miller, Kerry. "Pearl Ridge Students Net State Insect." Midweek, June 3, 2009. http://archives.midweek.com/content/zones/west_coverstory_article/pearl_ridge_students_net_state_insect/.

Pulelehua Project. iNaturalist. https://www.inaturalist.org/projects/pulelehua-project.

Pulelehua Project. University of Hawai'i at Mānoa. https://cms.ctahr.hawaii.edu/pulelehua/.

"Researchers need help saving the Kamehameha butterfly." University of Hawai'i News, February 17, 2014. https://www.youtube.com/watch?v=1a378SO6K2g.

"Successful Breeding and Release of Kamehameha Butterflies." Honolulu Zoo Society, December 28, 2017. https://honoluluzoo.org/successful-breeding-and-release-of-kamehameha-butterflies/.

Williams, Francis X. "The Kamehameha Butterfly, Vanessa tammeamea Esch" [misspelling of *tameamea*]. *Proceedings of the Hawaiian Entomological Society* 7, no. 1 (1928): 164–169. https://scholarspace.manoa.hawaii.edu/bitstream/10125/15763/PHES07_164-169.pdf.

Yuen, Nate. "Kamehameha Butterflies in the Koa Forests." Hawaiian Forest, July 31, 2013. http://hawaiianforest.com/wp/kamehameha-butterflies-in-the-koa-forests/.

Photograph Credits

map: Armita/Shutterstock; statue of King Kamehameha, māmaki plant inset, and butterflies at Honolulu Zoo: Harry Trumbore; butterfly egg, fifth instar caterpillar, female and male adult Kamehameha butterflies, and folded-wing butterfly: Dr. Will Haines; chrysalis, growing butterflies at insectary, māmaki plant, and koa tree: Cindy Trumbore; Pearl Ridge students: teacher Laura Brown, used with permission of Pearl Ridge Elementary School; Dr. Will Haines at insectary and butterfly born at insectary: Susan L. Roth; koa tree inset: David Eickhoff from Pearl City, Hawaii, USA.

Edited by Louise E. May
Designed by Christy Hale
Production by The Kids at Our House
The text is set in Body Grotesque and Asimov
The illustrations are rendered in paper and fabric collage
Manufactured in China by Jade Productions
Printed on paper from responsible sources
10 9 8 7 6 5 4 3 2 1
First Edition

Library of Congress Cataloging-in-Publication Data
Names: Roth, Susan L., author, illustrator. | Trumbore, Cindy, author.
Title: Butterfly for a king : saving Hawai'i's Kamehameha butterflies / Susan L. Roth and Cindy Trumbore ; collages by Susan L. Roth.
Description: New York : Lee & Low Books, 2021. | Includes bibliographical references. |
Audience: Ages 8-12 | Audience: Grades 4-6 | Summary: "A combined history of the Hawaiian Islands and the native Kamehameha butterfly up to and including current-day efforts of Hawai'i's Pulelehua Project, a group of professional and citizen scientists working to restore the butterfly's declining habitats and population. An Afterword with additional information, photographs, and source list is included"—Provided by publisher.
Identifiers: LCCN 2020013580 | ISBN 9781620149713 (hardcover)
Subjects: LCSH: Vanessa tameamea—Conservation—Hawaii—Juvenile literature.
Classification: LCC QL561.N9 R68 2020 | DDC 595.78/9—dc23
LC record available at https://lccn.loc.gov/2020013580